I0466863

THE CODE OF THE COSMOS

Billy Carson's Algorithmic Philosophy

Oteren.Fredrick

DISCLAIMER

The substance introduced in The Algorithmic Universe: Billy Carson's Investigation of Astronomical Codes is expected for enlightening and instructive purposes as it were. The perspectives and speculations examined in this book depend on the translations of Billy Carson and the writer's examination and union of related research. While endeavors have been made to guarantee the exactness and dependability of the data gave, this book doesn't ensure the culmination or rightness of the substance.

COPYRIGHT

TABLE OF CONTENT

INTRODUCTION

The universe has forever been a subject of interest, motivating endless speculations and methods of reasoning about its starting point, structure, and the regulations that oversee it. Among the bunch of scholars and visionaries who have tried to comprehend the universe, Billy Carson stands apart with his progressive viewpoint: the conviction that the universe is on a very basic level algorithmic in nature. "The Code of the Universe: Billy Carson's Algorithmic Way of thinking" digs into this captivating idea, investigating Carson's interesting understanding of the universe as a mind boggling, code-based substance.

Billy Carson, a wayfarer of old civic establishments, specialist of stowed away information, and supporter of the possibility that our world is organized by modern calculations, expresses an impression inciting focal point through which we can see presence. His work difficulties conventional logical and philosophical standards, proposing that the universe works much the same way to a tremendous, unpredictable PC program. In this presentation, we will give an outline of Carson's experience, the embodiment of his algorithmic way of thinking, and the meaning of investigating such a point of view in our journey to grasp the universe.

Billy Carson: A Concise Foundation

Billy Carson's excursion into the profundities of old information and present day science started with a significant interest in his general surroundings. His different vocation traverses different fields, including paleontology, space science, and computerized innovation, which aggregately illuminate his remarkable perspective. Carson is the organizer behind 4biddenknowledge Inc., an association devoted to revealing and spreading stowed away bits of insight about mankind's set of experiences, otherworldliness, and the universe. Through his books, narratives, and public talking commitment, Carson has propelled numerous to address laid out accounts and look for more profound comprehension.

Carson's investigation of antiquated texts, like the Emerald Tablets of Thoth and Sumerian cuneiform tablets, has driven him to infer that early civic establishments had progressed information about the idea of the real world. He contends that these old works contain references to modern calculations and codes that support the texture of the universe. By incorporating bits of knowledge from these messages with current logical disclosures, Carson presents a convincing case for his algorithmic universe hypothesis.

The Algorithmic Universe: An Outline

At the core of Billy Carson's way of thinking is the possibility that the universe is definitely not an irregular, tumultuous substance but instead an organized, algorithmic framework represented by exact codes and examples. This idea draws matches between the functions of the universe and the tasks of a PC, where complex calculations direct the way of behaving of different components. As per Carson, these calculations are answerable for the mind boggling request and concordance saw in nature, from the winding examples of universes to the fractal math of shorelines.

In Carson's view, everything in the universe — from the littlest particles to the biggest vast designs — follows explicit calculations. These calculations are much the same as the code that runs a PC program, deciding how components cooperate, develop, and capability. This viewpoint challenges the conventional logical view that stresses arbitrariness and likelihood as basic parts of the universe. All things considered, Carson's way of thinking recommends that there is a hidden request and reason to all things, encoded in the actual texture of the real world.

One of the most convincing parts of Carson's hypothesis is its interdisciplinary nature. By drawing associations between old insight, state of the art science, and computerized innovation, he offers an all encompassing system for figuring out the universe. This approach resounds with the developing interest in integrative and cross-disciplinary examination, which looks to connect holes between various fields of information and uncover brought together insights about presence.

The Meaning of an Algorithmic Viewpoint

Investigating the universe from the perspective of calculations has significant ramifications for different areas of information, including science, reasoning, and otherworldliness. From a logical stance, the algorithmic universe hypothesis urges scientists to look for more profound examples and designs hidden normal peculiarities. It moves researchers to look past superficial perceptions and uncover the codes that oversee the way of behaving of issue and energy. This viewpoint could prompt pivotal disclosures and advancements, especially in fields like quantum mechanics, computerized reasoning, and computational science.

Logically, Carson's algorithmic universe welcomes us to reevaluate how we might interpret reality and our place inside it. On the off chance that the universe is to be sure administered by calculations, our reality may not be essentially as irregular or purposeless as it at times shows up. All things considered, we could be important for a fantastic, savvy configuration, unpredictably woven into the texture of the universe. This thought reverberates with different philosophical and otherworldly practices that underline the interconnectedness and deliberateness, everything being equal.

According to an otherworldly viewpoint, the algorithmic universe lines up with the idea of a heavenly or clever maker who planned the universe with goal and accuracy. Numerous strict and otherworldly lessons discuss a higher request and reason behind the universe, frequently depicting it concerning heavenly knowledge or enormous cognizance. Carson's hypothesis gives a cutting edge, logically educated structure for understanding these old convictions, recommending that the calculations overseeing the universe are articulations of this higher knowledge.

Besides, Carson's algorithmic way of thinking has commonsense ramifications for how we carry on with our lives and communicate with our general surroundings. By perceiving the algorithmic idea of the real world, we can foster a more profound appreciation for the complicated examples and frameworks that support life. This mindfulness can motivate us to live more agreeably with nature, settle on additional educated choices, and seek after information and development with a feeling of direction and interest.

The Excursion Ahead

In "The Code of the Universe: Billy Carson's Algorithmic Way of thinking," we will leave on an excursion to investigate the different features of Carson's hypothesis, from its fundamental standards to its broad ramifications. We will dive into the antiquated texts and current logical disclosures that help his perspectives, analyze the examples and codes that swarm nature, and think about the connection among awareness and calculations. En route, we will address normal reactions and counterarguments, giving a reasonable and exhaustive comprehension of this intriguing viewpoint.

Eventually, this book plans to move perusers to ponder the idea of the universe and their place inside it. Whether you are a carefully prepared researcher, an inquisitive layman, or a profound searcher, Carson's algorithmic way of thinking expresses a new and impression inciting focal point through which to see presence. By investigating the code of the universe, we can open new bits of knowledge into the secrets of the universe and our own excursion through life.

CHAPTER 1:

The Foundations of Carson's Philosophy

Billy Carson's perspective, which sets that the universe works like a staggering, refined computation, depends on a supporting of various effects, experiences, and intensive investigation. Understanding these focal parts is pressing to making sense of the significance and broadness of Carson's algorithmic universe speculation. This part explores Carson's underlying life, his key effects, and the basic experiences that have formed his extraordinary perspective on the universe.

Early Effects and Inspirations

Billy Carson's advantage with the mysteries of the universe began in his life as a youth. Growing up, he was surrounded by an overflow of books and resources that lit his advantage in the world past customary acumen. His underlying receptiveness to science fiction, old-fashioned legends, and coherent composing planted the seeds for his later examinations concerning the possibility of this present reality. These beginning phases were separate by a ravenous hankering to grasp how things work, a quality that would describe his enduring excursion for data.

One of Carson's earliest and most tremendous effects was made by Erich von Däniken, particularly his book "Chariots of the Heavenly creatures?" which examines the opportunity of obsolete space adventurers visiting Earth. This book opened Carson's mind to the likelihood that old improvements could have had advanced data and headways quite far past what normal history suggests. Von Däniken's work spiced up Carson to dive into the examination of obsolete texts and doodads, searching for evidence of hid away data about the universe.

Another huge effect on Carson's thinking was the field of quantum mechanics. The uncommon and silly eccentricities saw at the quantum level, for instance, catch and superposition, enthralled him and drove him to investigate the fundamental thought of this present reality. The likelihood that particles could exist in various states simultaneously and be interconnected across colossal distances resonated with his creating confidence in an essential solicitation to the universe.

Key Experiences Trim His Points of view

All through his life, Billy Carson has had a couple of imperative experiences that have essentially shaped his points of view on the universe. One such experience was his examination of the Emerald Tablets of Thoth, an old plan of texts credited to the unbelievable figure Thoth, the Egyptian heavenly power of knowledge. Carson's examination of these tablets convinced him that they contained encoded data about the possibility of this present reality and the universe's algorithmic development.

Carson's developments to old archeological regions all around the planet further set his trust in the general data on early city foundations. Visiting places like the pyramids of Egypt, the enormous plans of Stonehenge, and the old city of Teotihuacan in Mexico, Carson was struck by the precision and multifaceted design of these turns of events. He speculated that such achievements of planning were possible just with a significant understanding of the mathematical and algorithmic principles that direct the universe.

Another weighty experience for Carson was his responsibility with current electronic advancement. As a tech darling and financial specialist, Carson lowered himself in the domain of PC programming and man-made cognizance. The equivalents between the action of PC estimations and the models found in nature developed his confidence in an algorithmic universe. He began to see the universe as a gigantic, insightful structure, comparable as a puzzling PC program, running on refined code.

Layout of His Past Work and Responsibilities

Billy Carson's trip into the algorithmic thought of the universe has driven him to commit to enormous responsibilities across various fields. He is the essayist of a couple of books that research old data, hidden away history, and the possibility of this present reality. His works habitually draw relationship between old messages, present day science, and significant examples, offering perusers a careful and integrative perspective on the universe.

Carson's affiliation, 4biddenknowledge Inc., has transformed into a phase for scattering his investigation and considerations. Through this affiliation, he has conveyed stories, worked with courses, and made educational substance highlighted uncovering the mysterious pieces of understanding of our world. His work has gathered a gigantic following, with numerous people finding inspiration in his far reaching method for managing sorting out the universe.

One of Carson's famous responsibilities is his assessment into sanctified estimation, which researches the mathematical models and shapes that underlie the development of the universe. He fights that these numerical designs, for instance, the Blossom of Life and the Fibonacci plan, are evidence of the algorithmic thought of this present reality. His work in this space has overcomed any hindrance between old knowledge and current science, exhibiting the way that eternal norms can prompt our contemporary understanding with respect to the universe.

Carson has moreover been a vocal advertiser for the joining of supernatural quality and science. He acknowledges that the algorithmic universe speculation gives a framework to obliging legitimate divulgences with significant illustrations, offering a bound together viewpoint on this present reality. This integrative technique has resounded with numerous people searching for a more significant cognizance of their spot in the universe.

The Presentation of the Algorithmic Universe Speculation

The likelihood that the universe works like an immense, sharp estimation didn't emerge for the present. It was the outcome of extensive stretches of study, assessment, and association of grouped fields of data. Carson's speculation depends on the explanation that the models and plans found in nature are not sporadic anyway are directed by unambiguous codes and estimations. These codes, he battles, are similarly as the programming vernaculars used in computer programming, coordinating the approach to acting and collaborations of all parts in the universe.

Carson's algorithmic universe theory sets that everything in presence, from subatomic particles to vast turns of events, observes precise mathematical guidelines. These guidelines are the "code" that underlies reality, pursuing solicitation and arrangement out of clear chaos. By understanding these estimations, Carson acknowledges we can open the insider realities of the universe and gain further pieces of information into the possibility of awareness, life, and presence itself.

CHAPTER 2:

The Universe as a Code

Billy Carson's algorithmic universe hypothesis sets that the universe works in much the same way to a huge, complex PC program, represented by exact codes and calculations. This part dives into the key standards of Carson's hypothesis, investigating the way in which calculations assume a focal part in organizing and directing the universe. We will analyze what calculations are, the means by which they manifest in the regular world, and the ramifications of review the universe through this computational focal point.

Figuring out Calculations

A calculation is a bunch of directions or rules intended to play out a particular errand or take care of a specific issue. With regards to software engineering, calculations are the foundation of programming programs, directing how information is handled, dissected, and changed. Calculations can go from basic cycles, such as arranging a rundown of numbers, to complex tasks, for example, foreseeing weather conditions or mimicking brain organizations.

In Carson's hypothesis, calculations are not bound to the domain of advanced innovation. All things considered, they are key rules that oversee the way of behaving of everything in the universe. As indicated by Carson, the very rationale and request that drive PC programs likewise support the laws of material science, the arrangement of worlds, and the development of life.

The Algorithmic Idea of Regular Peculiarities

One of the vital parts of Carson's hypothesis is the perception of algorithmic examples in nature. These examples recommend that normal peculiarities are not arbitrary yet keep explicit guidelines and designs. Here are a few models that show this idea:

1. Fractals: Fractals are intricate mathematical shapes that can be parted into parts, every one of which is a decreased scale duplicate of the entirety. This self-comparative example is clear in different normal developments, like the spreading of trees, the design of snowflakes, and the shoreline's rough edges. The numerical conditions that create fractals are similar to calculations, delivering many-sided and limitlessly nitty gritty designs.

2. The Fibonacci Sequence: The Fibonacci grouping is a progression of numbers where each number is the amount of the two going

before ones. This grouping shows up in numerous natural settings, like the plan of leaves on a stem, the example of seeds in a sunflower, and the twisting shells of mollusks. The repeat connection that characterizes the Fibonacci grouping is a straightforward calculation that appears in the normal world.

3. Cellular Automata: Cell automata are numerical models used to reenact complex frameworks and cycles. These models comprise of a lattice of cells, every one of which can be in one of a limited number of states. The condition of every phone changes as indicated by a bunch of rules in light of the conditions of its adjoining cells. Cell automata can produce mind boggling and dynamic examples, impersonating peculiarities like gem development and populace elements.

4. DNA and Hereditary Code: The design and capability of DNA are great representations of the algorithmic idea of life. DNA arrangements are basically natural calculations, encoding

directions for building and keeping up with living creatures. The cycles of record and interpretation, which convert hereditary data into proteins, observe exact guidelines much the same as PC calculations.

The Numerical Standards Behind the Universe

Carson's algorithmic universe hypothesis additionally stresses the significance of numerical standards in figuring out the universe. Math is many times portrayed as the language of the universe, and its standards are profoundly implanted in the texture of the real world. Here are a few key numerical ideas that help Carson's perspectives:

1. Chaos Theory: Confusion hypothesis concentrates on the way of behaving of dynamical frameworks that are exceptionally delicate to beginning circumstances. Little changes in these circumstances can prompt boundlessly various results, a peculiarity frequently alluded to as the "butterfly impact."

In spite of the clear haphazardness, tumultuous frameworks keep deterministic guidelines and calculations. Carson contends that disarray hypothesis uncovers the basic request and intricacy of the universe, represented by numerical regulations.

2. Symmetry and Gathering Theory: Balance assumes a urgent part in the laws of physical science and the construction of the universe. Bunch hypothesis, a part of math that concentrates on balances, makes sense of basic particles and powers. For instance, the Standard Model of molecule material science depends on bunch hypothesis to portray the communications between rudimentary particles. Carson considers balance and gathering hypothesis to be proof of the algorithmic idea of the real world, where explicit principles and examples direct the way of behaving of issue and energy.

3. Information Theory: Data hypothesis, created by Claude Shannon, measures the transmission, handling, and capacity of data. Carson draws

matches between data hypothesis and the algorithmic universe, proposing that the universe processes and communicates data as per explicit calculations. This viewpoint lines up with the possibility that the universe is a huge, wise framework, continually encoding and deciphering data.

Ramifications of the Algorithmic Universe

Seeing the universe as an algorithmic element has significant ramifications for different fields of information. Here are a portion of the possible effects of this viewpoint:

1. Scientific Research: The algorithmic universe hypothesis urges researchers to look for more profound examples and designs fundamental regular peculiarities. By understanding the calculations that administer the universe, specialists can foster new models and speculations to make sense of the way of behaving of issue and energy. This approach could prompt leap forwards in fields like

quantum mechanics, cosmology, and computational science.

2. Philosophy: The possibility of an algorithmic universe challenges conventional philosophical perspectives on irregularity and determinism. On the off chance that the universe works as per exact calculations, occasions and cycles are not arbitrary yet adhere to explicit guidelines. This point of view brings up issues about choice, causality, and the idea of the real world, inciting rationalists to reevaluate how they might interpret presence.

3. Spirituality: Carson's algorithmic universe hypothesis resounds with different otherworldly lessons that underscore the request and reason for the universe. Numerous otherworldly practices depict the universe as a keen, interconnected framework, directed by divine standards. The algorithmic viewpoint gives a cutting edge, logically educated structure for understanding these convictions, recommending

that the calculations overseeing the universe are articulations of higher knowledge.

4. Technology: The equals between the algorithmic idea of the universe and advanced innovation open up additional opportunities for development. By concentrating on the calculations that administer regular peculiarities, researchers and designers can foster new advances that copy or saddle these standards. This approach could prompt progressions in man-made brainpower, quantum registering, and bioengineering.

CHAPTER 3:

Patterns in Nature

Cell Automata: Reenacting Regular Cycles

Cell automata are numerical models used to mimic complex frameworks and cycles. They comprise of a network of cells, every one of which can be in one of a limited number of states. The condition of every phone changes as indicated by a bunch of rules in light of the conditions of its adjoining cells. Cell automata can create mind boggling and dynamic examples, copying regular peculiarities. Key models include:

1. Conway's Down of Life: Created by mathematician John Conway, this cell machine shows the way that straightforward standards can prompt complex ways of behaving. The Round of Life reenacts populace elements, showing how examples arise, develop, and now and again balance out or vanish.

2. Crystal Growth: Cell automata models can imitate the course of gem development, where

molecules orchestrate themselves into organized, rehashing designs.

3. Biological Systems: Cell automata have been utilized to display different natural cycles, like the advancement of creatures, the spread of infections, and the working of brain organizations.

The capacity of cell automata to recreate regular cycles with basic calculations upholds Carson's view that the universe works on central standards similar to PC code.

Consecrated Math: The Language of the Universe

Hallowed math alludes to the investigation of mathematical shapes and examples that are accepted to have otherworldly importance and mirror the fundamental request of the universe. These shapes frequently relate to numerical standards and show up in different normal and man-made structures. Models include:

1. The Bloom of Life: This multifaceted example comprises of numerous, equally separated, covering circles organized to frame a blossom like plan. It is tracked down in old craftsmanship and design and is remembered to address the interconnectedness of all life.

2. The Brilliant Ratio: The brilliant proportion, roughly 1.618, is a numerical consistent that shows up in different normal and imaginative settings. It is related with stylishly satisfying extents and can be tracked down in the twistings of shells, the fanning of trees, and the human body.

3. Platonic Solids: These are the five customary polyhedra (tetrahedron, 3D shape, octahedron, dodecahedron, and icosahedron) that are exceptionally balanced and have been contemplated since relic. They show up in normal structures like precious stones and infections.

Sacrosanct calculation represents how numerical standards and mathematical examples support the design of the universe, building up the possibility of an algorithmic universe.

Contextual analyses and Genuine Models

To additionally show the algorithmic idea of the universe, let us look at a couple of contextual analyses and certifiable models that feature the presence of examples and codes in nature:

1. The Nautilus Shell: The nautilus shell is an exemplary illustration of the logarithmic winding, an example that can be produced utilizing a straightforward numerical calculation. The winding's development follows the brilliant proportion, making a shape that is both productive and stylishly satisfying.

2. The Human Genome: The sequencing of the human genome uncovered the mind boggling calculations that administer the turn of events

and working of living life forms. The hereditary code, made out of successions of nucleotides, goes about as a bunch of guidelines for building proteins and controlling natural cycles.

3. Weather Patterns: The calculations hidden weather conditions, like the arrangement of typhoons and twisters, exhibit the interaction among disorder and request. Meteorologists utilize numerical models to foresee weather conditions in view of these basic standards.

Ramifications of Algorithmic Examples in Nature

The presence of algorithmic examples in nature has significant ramifications for how we might interpret the universe:

1. Scientific Research: Perceiving the algorithmic idea of normal peculiarities can prompt new logical disclosures and mechanical advancements. By revealing the calculations that

administer the universe, analysts can foster more exact models and recreations.

2. Philosophical Inquiry: The presence of these examples challenges how we might interpret arbitrariness and determinism. Assuming that the universe works on exact calculations, it brings up issues about freedom of thought, causality, and the idea of the real world.

3. Spiritual Understanding: The arrangement of hallowed calculation with logical standards proposes a more profound association among otherworldliness and science. This point of view can overcome any barrier between various perspectives and give a comprehensive comprehension of presence.

4. Technological Innovation: Concentrating on the calculations of nature can rouse new advancements that emulate regular cycles. Biomimicry, the plan of frameworks and materials demonstrated after organic elements, is

one such field that advantages from grasping regular calculations.

CHAPTER 4:

Consciousness and the Algorithmic Universe

Billy Carson's algorithmic universe hypothesis stretches out past the actual world, placing that awareness itself is a fundamental piece of this stupendous, shrewd framework. This part digs into the connection among awareness and the algorithmic idea of the universe, investigating how our brains may be impacted by, and even associate with, the essential codes that oversee reality. We will inspect speculations of cognizance, the idea of an all inclusive psyche, and the ramifications of review awareness as an algorithmic interaction.

The Idea of Cognizance

Cognizance is perhaps of the most significant and slippery peculiarity known to man. It envelops our contemplations, feelings, discernments, and mindfulness. Regardless of broad review, the real essence of cognizance stays a secret. Different hypotheses endeavor to make sense of its starting point and capability:

1. Materialist Theories: Realist speculations declare that cognizance emerges simply from actual cycles in the cerebrum. As indicated by this view, brain organizations and electrochemical signs create our cognizant experience.

2. Dualist Theories: Dualist hypotheses recommend that cognizance is independent from the actual body and mind. This viewpoint recommends that a non-material psyche or soul

connects with the actual mind to deliver cognizant experience.

3. Panpsychism: Panpsychism sets that cognizance is a major property of the universe, present in all make a difference somewhat. As indicated by this view, even rudimentary particles have a type of simple cognizance.

4. Integrated Data Hypothesis (IIT): IIT proposes that cognizance emerges from the combination of data inside a framework. The more coordinated and complex the data, the higher the degree of awareness.

Cognizance as an Algorithmic Cycle

Carson's algorithmic universe hypothesis lines up with the possibility that cognizance could be perceived as an algorithmic cycle. This viewpoint proposes that cognizance rises out of the cooperation of essential codes and calculations that oversee the universe. Here are a few central issues to consider:

1. Neural Organizations and Computation: The human mind should be visible as an exceptionally intricate brain organization, handling tremendous measures of data through interconnected neurons. This computational view lines up with the idea of calculations, where explicit standards and examples direct the mind's working.

2. Patterns of Thought: Our considerations and discernments frequently follow conspicuous examples, which can be compared to algorithmic cycles. For instance, navigation includes gauging choices in view of a bunch of standards, like how a PC calculation assesses information to arrive at a resolution.

3. Conscious Experience and Data Processing: Assuming cognizance emerges from the handling of data, as proposed by IIT, it suggests that our cognizant experience is administered by hidden calculations. These calculations incorporate tactile information sources,

recollections, and mental cycles to make an intelligent mindfulness.

4. Quantum Consciousness: A few hypotheses recommend that cognizance is connected to quantum processes in the cerebrum. Quantum mechanics, with its intrinsic probabilistic and interconnected nature, could give a system to figuring out the algorithmic premise of cognizance.

The Widespread Brain

Carson's hypothesis likewise engages the thought of an all inclusive brain, a sweeping cognizance that penetrates the whole universe. This idea lines up with specific profound and philosophical customs that view the universe as a living, cognizant substance. Key parts of this thought include:

1. Cosmic Consciousness: The possibility that the actual universe has a type of cognizance is

reverberated in different otherworldly and strict practices. This grandiose cognizance is viewed as the wellspring of all knowledge and request in the universe.

2. The Akashic Records: In numerous obscure customs, the Akashic Records are accepted to be a grandiose store of all information and encounters. This idea proposes that all data is put away in the texture of the universe, open through a higher condition of cognizance.

3. Non-Region and Interconnectedness: Quantum ensnarement shows the way that particles can be interconnected across huge distances, affecting each other promptly. This non-territory suggests a more profound degree of interconnectedness in the universe, conceivably stretching out to cognizance itself.

4. Holographic Principle: The holographic standard sets that the whole universe can be depicted as a two-layered data structure projected onto a three-layered space. This

thought upholds the idea that cognizance and data are crucial parts of the real world, encoded in the texture of the universe.

Ramifications of Algorithmic Awareness

Seeing awareness as an algorithmic interaction has significant ramifications for different fields of information and human experience:

1. Neuroscience and AI: Understanding cognizance as a calculation could alter neuroscience and computerized reasoning. Specialists could foster more refined models of cerebrum capability and make artificial intelligence frameworks that imitate human-like awareness.

2. Philosophy and Ethics: The algorithmic idea of awareness challenges conventional philosophical perspectives on the brain body issue and unrestrained choice. It brings up issues

about the idea of selfhood, character, and moral obligation.

3. Spirituality and Mysticism: The idea of a general brain overcomes any barrier among science and otherworldliness. It gives a system to figuring out enchanted encounters, instinct, and the feeling of interconnectedness that numerous profound customs depict.

4. Mental Wellbeing and Well-being: Review cognizance as an algorithmic cycle can prompt new methodologies in emotional well-being treatment. Methods that change figured designs, like mental social treatment, should have been visible as reinventing the calculations that underlie mental cycles.

5. Human Potential: Assuming awareness works on algorithmic standards, it proposes that we can advance and upgrade our intellectual abilities. Practices like contemplation, care, and neurofeedback could be seen as ways of refining the calculations of the brain.

Contextual investigations and Models

To outline the algorithmic idea of cognizance, let us inspect a couple of contextual investigations and models that feature the interaction among psyche and calculation:

1. Meditation and Mind Waves: Concentrates on contemplation have demonstrated the way that it can adjust cerebrum wave designs, advancing conditions of unwinding and uplifted mindfulness. These progressions in mind movement mirror the reconstructing of brain calculations, prompting adjusted conditions of cognizance.

2. Synesthesia: Synesthesia is a condition where excitement of one tangible pathway prompts programmed, compulsory encounters in another pathway. This peculiarity proposes that tangible handling follows explicit calculations, which can incidentally meet and make extraordinary perceptual encounters.

3. Artificial Intelligence: Advances in artificial intelligence, especially in brain organizations and AI, give bits of knowledge into the algorithmic idea of awareness. Artificial intelligence frameworks that copy human insight show the way that intricate calculations can create shrewd way of behaving and navigation.

4. Psychedelic Experiences: Hallucinogenics can significantly change cognizance, uncovering the pliancy of the psyche. These substances could briefly rework the mind's calculations, prompting changed discernments, bits of knowledge, and a feeling of interconnectedness.

CHAPTER 5:

The Intersection of Science and Spirituality

Billy Carson's algorithmic universe hypothesis gives a remarkable extension among science and otherworldliness, offering a system to accommodate these frequently different fields. This section investigates how seeing the universe as a complicated calculation can orchestrate logical request with profound insight, revealing insight into key inquiries concerning presence, reason, and the idea of the real world. We will analyze key logical ideas and profound lessons, showing how they meet and complete one another.

Logical Request and the Algorithmic Universe

Science is driven by the mission to figure out the regular world through perception, trial and error, and levelheaded investigation. The algorithmic universe hypothesis offers another focal point through which to decipher logical disclosures and peculiarities, accentuating the fundamental request and knowledge of the universe. Here are a few logical ideas that line up with Carson's hypothesis:

1. The Laws of Physics: The laws of physical science, like gravity, electromagnetism, and quantum mechanics, should be visible as the essential calculations administering the way of behaving of issue and energy. These regulations are exact and predictable, mirroring the numerical request fundamental the universe.

2. Information Theory: Data hypothesis, which concentrates on the transmission, handling, and capacity of data, upholds the possibility that the universe works on enlightening standards. The idea of the universe as a huge data handling

framework lines up with the algorithmic viewpoint.

3. Complex Frameworks and Emergence: In complex frameworks, basic standards can prompt emanant ways of behaving and structures. This standard is obvious in natural development, atmospheric conditions, and, surprisingly, social frameworks. The rise of intricacy from straightforward calculations repeats Carson's perspective on the universe.

4. Simulation Hypothesis: The recreation speculation, which recommends that reality could be a modern virtual experience, matches the algorithmic universe hypothesis. In the event that the universe works like a reproduction, it follows explicit codes and calculations.

Profound Insight and the Algorithmic Universe

Otherworldly lessons frequently depict the universe as an interconnected, clever substance. These lessons underline the solidarity of all presence, the presence of a higher request, and the meaning of inward encounters. Carson's hypothesis resounds with numerous otherworldly ideas, making a shared belief for discourse. Here are a few key profound thoughts that line up with the algorithmic point of view:

1. Unity and Interconnectedness: Numerous otherworldly practices instruct that all life is interconnected and part of a more noteworthy entirety. This thought lines up with the algorithmic universe hypothesis, which sets that all that in the universe is represented by similar crucial standards and codes.

2. Divine Intelligence: The idea of a heavenly insight or grandiose brain that overruns the universe is fundamental to numerous otherworldly convictions. Carson's hypothesis proposes that the calculations overseeing the universe are articulations of this higher knowledge.

3. Mystical Experiences: Enchanted encounters frequently include a feeling of solidarity, greatness, and direct information on the universe's basic request. These encounters can be deciphered as minutes when people access the more profound calculations of the real world.

4. Sacred Geometry: Sacrosanct calculation, which investigates the otherworldly meaning of mathematical examples, lines up with the possibility that the universe is organized by basic numerical standards. These examples are viewed as the outlines of creation.

Contextual analyses and Genuine Models

To outline the convergence of science and otherworldliness from the perspective of the algorithmic universe, we should look at a couple of contextual investigations and certifiable models:

1. Meditation and Neuroscience: Exploration on reflection has demonstrated the way that it can

adjust mind movement and advance prosperity. These progressions mirror the reconstructing of brain calculations, featuring the crossing point of otherworldly practices and logical comprehension.

2. Quantum Mechanics and Mysticism: Quantum mechanics uncovers a reality that is probabilistic, interconnected, and impacted by perception. These standards resound with supernatural lessons that underscore the liquid and interconnected nature of the real world.

3. Fractals in Nature and Art: The presence of fractal designs in normal arrangements and holy workmanship outlines the assembly of logical and profound points of view. These examples mirror the algorithmic idea of the universe and its intrinsic request.

4. Holistic Wellbeing Practices: Comprehensive wellbeing rehearses, for example, needle therapy and energy recuperating, frequently accentuate the progression of energy and the

interconnectedness of the body and brain. These ideas line up with the possibility of an algorithmic universe, where wellbeing and prosperity are administered by hidden codes.

Suggestions for Science and Otherworldliness

Seeing the universe as an algorithmic substance has significant ramifications for both science and otherworldliness:

1. Integration of Knowledge: The algorithmic universe hypothesis supports the combination of logical and otherworldly information. By perceiving the normal standards hidden the two spaces, we can foster a more thorough comprehension of the real world.

2. Ethical and Moral Perspectives: Understanding the universe as an interconnected framework administered by calculations can illuminate our moral and moral structures. It

stresses the significance of congruity, equilibrium, and regard for all types of life.

3. Innovative Research: The algorithmic viewpoint can rouse new roads of logical exploration and mechanical advancement. By concentrating on the major codes of the universe, we can foster trend setting innovations and develop how we might interpret normal peculiarities.

4. Personal and Aggregate Transformation: Embracing the algorithmic idea of the universe can prompt individual and aggregate change. It urges us to develop care, sympathy, and a feeling of solidarity with all presence.

The Assembly of Science and Otherworldliness

The algorithmic universe hypothesis represents the assembly of logical and otherworldly points of view. By review the universe as a mind boggling, wise framework represented by crucial codes, we can see the value in the significant interconnectedness, everything being equal. This intermingling offers an all encompassing comprehension of presence, where science and otherworldliness supplement and improve one another.

CHAPTER 6:

Technology and the Algorithmic Universe

Billy Carson's hypothesis of the algorithmic universe reaches out past grasping regular peculiarities and cognizance, giving a structure to mechanical progression and development. This section investigates how seeing the universe as a mind boggling calculation can impact the improvement of innovations, especially in the fields of man-made brainpower, quantum figuring, and biotechnology. By tackling the standards basic the algorithmic universe, we can make advances that are more proficient, smart, and agreeable with the regular world.

Computerized reasoning and AI

Computerized reasoning (artificial intelligence) and AI are at the bleeding edge of mechanical development. These fields include making calculations that permit machines to gain from information and decide. Seeing the universe as an algorithmic substance can offer new bits of knowledge into man-made intelligence improvement:

1. Neural Networks: Brain organizations, enlivened by the human mind, comprise of interconnected hubs that cycle data in a way like natural neurons. By concentrating on the algorithmic standards of brain organizations, we can foster computer based intelligence frameworks that mirror human perception all the more intently.

2. Deep Learning: Profound learning includes preparing multifaceted brain networks on huge datasets to perceive examples and make forecasts. Understanding the algorithmic idea of the universe can assist with refining profound learning strategies, making them more strong and equipped for taking care of perplexing errands.

3. Natural Language Handling (NLP): NLP centers around empowering machines to comprehend and produce human language. By applying the standards of the algorithmic universe, we can make NLP frameworks that are more natural and equipped for nuanced understanding.

4. Ethical AI: Review simulated intelligence advancement from the perspective of the algorithmic universe accentuates the significance of making moral and dependable man-made intelligence. This viewpoint supports the plan of computer based intelligence frameworks that

line up with major standards of concordance, equilibrium, and regard for all types of life.

Quantum Registering

Quantum figuring addresses a critical jump in computational power and effectiveness, utilizing the standards of quantum mechanics. The algorithmic universe hypothesis offers a system for understanding and propelling quantum figuring:

1. Quantum Algorithms: Quantum calculations exploit the standards of superposition and ensnarement to tackle issues more productively than traditional calculations. By concentrating on the algorithmic idea of the universe, specialists can foster new quantum calculations that address complex difficulties.

2. Simulation of Normal Processes: Quantum PCs can possibly recreate regular cycles with high exactness. This capacity lines up with the possibility that the universe works on essential

calculations, permitting us to profoundly demonstrate and figure out these cycles more.

3. Cryptography: Quantum cryptography use the standards of quantum mechanics to make secure correspondence frameworks. The algorithmic universe hypothesis upholds the advancement of powerful cryptographic conventions that are impervious to hacking.

4. Optimization Problems: Quantum figuring succeeds at tackling streamlining issues, which include tracking down the best arrangement from countless potential outcomes. Understanding the algorithmic standards of the universe can improve our capacity to proficiently handle these issues.

Biotechnology and Hereditary Designing

Biotechnology and hereditary designing include controlling organic frameworks to foster new clinical medicines, farming practices, and ecological arrangements. The algorithmic universe hypothesis gives a structure to propelling these fields:

1. CRISPR and Quality Editing: CRISPR innovation takes into account exact altering of hereditary material, much the same as changing the calculations that oversee natural cycles. By understanding the algorithmic idea of hereditary qualities, we can foster more designated and powerful quality treatments.

2. Synthetic Biology: Manufactured science includes planning and developing new organic parts, gadgets, and frameworks. Seeing science from the perspective of the algorithmic universe can motivate imaginative ways to deal with making engineered living beings that carry out unambiguous roles.

3. Personalized Medicine: Customized medication tailors clinical medicines to individual hereditary profiles. By applying the standards of the algorithmic universe, we can foster more exact and viable medicines that line

up with every individual's remarkable organic calculations.

4. Environmental Biotechnology: Ecological biotechnology utilizes natural cycles to address natural difficulties, for example, contamination and environmental change. Understanding the algorithmic idea of biological systems can improve our capacity to foster feasible and compelling arrangements.

Contextual   investigations   and   Genuine Applications

To outline the effect of the algorithmic universe hypothesis on innovation, we should look at a couple of contextual investigations and genuine applications:

1. AlphaGo and Support Learning: AlphaGo, an artificial intelligence created by DeepMind, utilizes support learning calculations to

dominate the round of Go. This artificial intelligence framework's prosperity shows the force of calculations in taking care of complicated issues and impersonating human key reasoning.

2. IBM's Quantum Computer: IBM's quantum PC, Q Framework One, addresses an achievement in quantum registering. By utilizing quantum calculations, this framework can perform calculations that are infeasible for traditional PCs, mirroring the capability of the algorithmic universe.

3. CRISPR-Cas9 Quality Editing: The CRISPR-Cas9 framework takes into account exact hereditary adjustments, offering additional opportunities for treating hereditary issues. This innovation represents the utilization of algorithmic standards to control organic frameworks.

4. Biodegradable Plastics: Specialists are creating biodegradable plastics utilizing natural

cycles and materials. These advancements line up with the algorithmic universe hypothesis, advancing practical arrangements that blend with regular frameworks.

Suggestions for Future Innovations

Seeing the universe as an algorithmic element has significant ramifications for future innovative turn of events:

1. Sustainable Technologies: The algorithmic point of view supports the improvement of advancements that are reasonable and as one with regular frameworks. This approach advances the production of eco-accommodating arrangements that address worldwide difficulties.

2. Human-Machine Collaboration: Understanding the algorithmic standards hidden both human insight and machine knowledge can upgrade human-machine cooperation. This

cooperative energy can prompt more productive and inventive critical thinking.

3. Ethical Considerations: The algorithmic universe hypothesis accentuates the significance of moral contemplations in mechanical turn of events. By adjusting innovation to major standards of equilibrium and regard, we can guarantee that developments benefit all types of life.

4. Holistic Innovation: Embracing the algorithmic idea of the universe urges a comprehensive way to deal with development, coordinating bits of knowledge from science, otherworldliness, and reasoning. This viewpoint cultivates a more profound comprehension of the interconnectedness, everything being equal.

CHAPTER 7:

Health and Well-being in the Algorithmic Universe

Billy Carson's algorithmic universe hypothesis reaches out into the domains of wellbeing and prosperity, recommending that our physical and mental states are administered by hidden calculations. This part investigates how understanding these calculations can prompt creative ways to deal with medical services, emotional well-being, and generally speaking prosperity. By lining up with the central codes of the universe, we can streamline our wellbeing and accomplish a condition of equilibrium and congruity.

The Algorithmic Idea of the Human Body

The human body is a perplexing framework that works as per unpredictable organic calculations. These calculations administer everything from cell cycles to the working of organs and frameworks. Understanding the algorithmic premise of the human body can improve clinical practices and wellbeing mediations:

1. Genetics and Epigenetics: Our hereditary code is a bunch of guidelines that direct the turn of events and working of our bodies. Epigenetics, which concentrates on how natural elements impact quality articulation, demonstrates the way that these guidelines can be adjusted. By figuring out the calculations of hereditary qualities and epigenetics, we can foster designated medicines and preventive measures.

2. Metabolism and Homeostasis: Metabolic cycles and the support of homeostasis are administered by biochemical calculations. These cycles guarantee that our bodies capability ideally, directing variables, for example, temperature, pH equilibrium, and energy levels.

3. Immune System: The resistant framework works through a complicated arrangement of calculations that distinguish and answer microorganisms. By concentrating on these calculations, we can upgrade our capacity to battle illnesses and foster compelling immunizations and treatments.

4. Neuroplasticity: The cerebrum's capacity to rearrange itself by shaping new brain associations, known as brain adaptability, is directed by fundamental calculations. Understanding these calculations can help in creating medicines for neurological issues and advancing mental wellbeing.

Emotional well-being and Cognizance

Emotional well-being is an essential part of in general prosperity, and it also can be grasped from the perspective of algorithmic cycles. By looking at the calculations that oversee our considerations, feelings, and ways of behaving, we can foster systems to work on psychological wellness and accomplish a condition of inward equilibrium:

1. Cognitive-Social Treatment (CBT): CBT depends on the possibility that our considerations and ways of behaving are interlinked and can be altered through unambiguous calculations. By distinguishing and changing negative idea designs, CBT assists people with accomplishing better psychological well-being.

2. Mindfulness and Meditation: Practices, for example, care and reflection include preparing the psyche to concentrate and accomplish a condition of quiet. These practices should be visible as strategies for reconstructing the cerebrum's calculations to advance unwinding and mental clearness.

3. Neurofeedback: Neurofeedback includes observing mind movement and giving criticism to assist people with managing their cerebrum

capability. This strategy use the mind's calculations to further develop psychological well-being conditions like uneasiness, despondency, and ADHD.

4. Psychedelic Therapy: Arising research recommends that hallucinogenics can adjust the cerebrum's calculations, prompting significant changes in cognizance and psychological well-being. These substances show guarantee in treating conditions like PTSD, discouragement, and enslavement.

Integrative Wellbeing Approaches

Integrative wellbeing approaches consolidate regular medication with correlative treatments, perceiving the significance of treating the entire individual. The algorithmic universe hypothesis upholds this all encompassing perspective, underscoring the interconnectedness of psyche, body, and soul:

1. Functional Medicine: Practical medication centers around distinguishing and tending to the main drivers of illness, instead of just treating side effects. This approach lines up with the algorithmic point of view, trying to comprehend and adjust the hidden organic calculations.

2. Ayurveda and Conventional Chinese Medication (TCM): Both Ayurveda and TCM accentuate equilibrium and agreement inside the body. These antiquated frameworks of medication depend on the comprehension of normal calculations and the progression of energy, offering important experiences into all encompassing wellbeing.

3. Nutrition and Stomach Health: Nourishment assumes a crucial part in our general wellbeing, impacting the calculations that oversee basicphysical processes. Understanding the connection between diet, stomach microbiota, and wellbeing can prompt customized healthful techniques that advance prosperity.

4. Energy Healing: Practices like Reiki, needle therapy, and Qi Gong include controlling the body's energy stream to reestablish harmony. These practices line up with the algorithmic universe hypothesis, accentuating the significance of energy designs and their effect on wellbeing.

Contextual analyses and Certifiable Applications

To outline the effect of the algorithmic universe hypothesis on wellbeing and prosperity, we should inspect a couple of contextual investigations and true applications:

1. Personalized Medicine: Advances in genomics and information examination have prompted the improvement of customized medication, which tailors medicines to a person's hereditary profile. This approach represents the use of algorithmic standards to enhance medical services.

2. Telemedicine and Computerized Health: Telemedicine and advanced wellbeing innovations use calculations to remotely screen and oversee wellbeing. These developments give open and effective medical services arrangements, especially in underserved regions.

3. Mental Wellbeing Apps: Portable applications that offer care works out, mental social methods, and mind-set following are turning out to be progressively famous. These apparatuses influence calculations to help psychological wellness and prosperity.

4. Biofeedback Devices: Gadgets that give continuous criticism on physiological boundaries, for example, pulse and breathing, assist people with managing their importantphysical processes. These gadgets apply algorithmic standards to advance unwinding and stress the board.

Suggestions for Wellbeing and Prosperity

Seeing wellbeing and prosperity from the perspective of the algorithmic universe hypothesis has significant ramifications:

1. Preventive Wellbeing Care: Understanding the calculations that administer our bodies can prompt more compelling preventive wellbeing measures. By distinguishing risk factors and early indications of illness, we can intercede before medical problems become serious.

2. Personal Empowerment: The algorithmic viewpoint engages people to assume responsibility for their wellbeing. By getting it and changing the calculations that impact their prosperity, individuals can settle on informed decisions and embrace better ways of life.

3. Holistic Treatments: The incorporation of ordinary and corresponding treatments can give more complete and powerful medicines. By taking into account the entire individual, instead of just side effects, we can accomplish better wellbeing results.

4. Innovative Therapies: The algorithmic universe hypothesis can motivate the improvement of new treatments and intercessions. By utilizing how we might interpret organic calculations, we can make inventive medicines for an extensive variety of ailments.

CHAPTER 8:

Exploring the Algorithmic Universe Through the Arts and Creativity

Billy Carson's algorithmic universe hypothesis impacts science and innovation as well as expands its venture into the domains of craftsmanship and inventiveness. This section investigates how understanding the universe as a mind boggling calculation can move imaginative articulation, upgrade inventive flows, and change our enthusiasm for workmanship. By inspecting the interchange among calculations and imaginative undertakings, we can uncover new components of innovativeness and articulation.

The Algorithmic Idea of Craftsmanship

Craftsmanship and imagination frequently include examples, designs, and rules, which should be visible as types of calculations. The algorithmic universe hypothesis gives a system to figuring out these parts of craftsmanship:

1. Mathematical Examples in Art: Numerous creative works consolidate numerical examples, for example, the Fibonacci grouping, fractals, and evenness. These examples mirror the basic calculations that administer both normal and creative structures.

2. Generative Art: Generative craftsmanship utilizes calculations to make visual or hear-able works. Specialists utilize code to create examples, shapes, and sounds, bringing about remarkable and complex organizations. This approach features the immediate use of algorithmic standards in innovative flows.

3. Algorithmic Composition: In music, algorithmic creation includes utilizing calculations to produce melodic designs and groupings. Writers can make intricate and imaginative pieces by programming rules and examples into their sytheses.

4. Digital Craftsmanship and AI: Computerized workmanship and simulated intelligence created

workmanship influence calculations to deliver new types of imaginative articulation. Man-made intelligence frameworks prepared on enormous datasets can produce fine art that reflects both human inventiveness and computational standards.

Inventiveness as an Algorithmic Cycle

Inventiveness includes creating clever thoughts and arrangements, frequently directed by basic cycles and designs. The algorithmic viewpoint offers bits of knowledge into how imagination capabilities:

1. Cognitive Algorithms: The innovative flow can be seen as a progression of mental calculations that include producing, assessing, and refining thoughts. Understanding these calculations can upgrade our capacity to encourage imagination and development.

2. Pattern Recognition: Innovativeness frequently includes perceiving and taking

advantage of examples. By understanding the calculations that support design acknowledgment, we can foster techniques to upgrade innovative reasoning and critical thinking.

3. Improvisation and Adaptation: Extemporization in music, dance, and different expressions should be visible as the utilization of adaptable calculations that adjust to evolving conditions. This point of view assists us with understanding how imagination develops in powerful conditions.

4. Collaborative Creativity: Cooperative innovativeness includes the communication of various people, each contributing their one of a kind viewpoints and abilities. The algorithmic point of view features how aggregate cycles can create inventive results through the interaction of different commitments.

Contextual investigations and Genuine Models

To represent the effect of the algorithmic universe hypothesis on craftsmanship and imagination, we should inspect a couple of contextual investigations and certifiable models:

1. Algorithmic Workmanship by Casey Reas: Casey Reas, one of the makers of Handling (a programming language for craftsmen), utilizes calculations to create visual craftsmanship. His works show the way that computational cycles can create stylishly convincing and complex pictures.

2. AI Craftsmanship by DeepArt.io: DeepArt.io utilizes brain organizations to make workmanship in light of info pictures and imaginative styles. The subsequent works of art mix human imagination with algorithmic cycles, displaying the capability of man-made intelligence in creative articulation.

3. Fractal Craftsmanship by Benoît B. Mandelbrot: Benoît B. Mandelbrot's investigation of fractals uncovered the

multifaceted examples that rise out of basic numerical calculations. Fractal craftsmanship, roused by Mandelbrot's work, represents the excellence of algorithmic designs in visual craftsmanship.

4. Generative Music by Brian Eno: Brian Eno's generative music projects use calculations to make advancing melodic organizations. Eno's work represents the way that calculations can deliver creative and dynamic soundscapes.

Suggestions for Imaginative Articulation

The algorithmic universe hypothesis has critical ramifications for imaginative articulation and inventiveness:

1. New Types of Art: The coordination of calculations and innovation into workmanship opens up additional opportunities for inventive articulation. Specialists can investigate novel

structures and procedures that push the limits of conventional craftsmanship.

2. Enhanced Creativity: Understanding the algorithmic standards fundamental imagination can prompt new strategies for improving creative cycles. Methods, for example, algorithmic conceptualizing and design acknowledgment can animate innovative reasoning.

3. Cross-Disciplinary Innovation: The convergence of workmanship, science, and innovation encourages cross-disciplinary advancement. Specialists and researchers cooperating can make momentous works that mirror the intricacy and excellence of the algorithmic universe.

4. Expanded Appreciation: The algorithmic point of view can extend our enthusiasm for workmanship by uncovering the basic designs and examples that add to its excellence. This understanding can improve our commitment

with imaginative works and their innovative flows.

The Fate of Craftsmanship and Imagination

As we keep on investigating the algorithmic universe hypothesis, the fate of workmanship and innovativeness holds invigorating conceivable outcomes:

1. Interactive Art: Advances in innovation and calculations will empower the making of intelligent workmanship that answers watchers' activities and data sources. This powerful commitment can make vivid and customized creative encounters.

2. AI-Helped Creativity: artificial intelligence devices and calculations will progressively help craftsmen in their innovative flows, giving new roads to trial and error and advancement. These apparatuses can upgrade imaginative abilities and open new skylines for innovative investigation.

3. Algorithmic Aesthetics: The investigation of algorithmic style will keep on developing, uncovering new bits of knowledge into the connection between arithmetic, calculations, and creative articulation. This investigation will extend how we might interpret excellence and innovativeness.

4. Global Collaboration: The worldwide idea of computerized and algorithmic craftsmanship will work with cooperation between specialists from different foundations and societies. This diverse trade will improve the imaginative scene and advance a more noteworthy enthusiasm for worldwide inventiveness.

CHAPTER 9:

The Algorithmic Universe in Philosophy and Ethics

Billy Carson's algorithmic universe hypothesis offers significant bits of knowledge into theory and morals, testing conventional viewpoints and proposing another system for grasping presence, ethical quality, and human way of behaving. This section investigates how the idea of the universe as a perplexing calculation impacts philosophical idea and moral contemplations, looking at its suggestions for how we might interpret reality, values, and navigation.

Philosophical Ramifications of the Algorithmic Universe

The algorithmic universe hypothesis challenges conventional philosophical thoughts by proposing that the universe works as per key codes and calculations. This viewpoint has huge ramifications for different parts of theory:

1. Ontology and Metaphysics: Philosophy, the investigation of being and presence, is impacted by the possibility that the universe is administered by calculations. This viewpoint proposes that the truth isn't irregular however follows explicit, deterministic examples. Supernatural inquiries concerning the idea of the real world, causality, and the idea of presence are rethought inside this unique situation.

2. Epistemology: Epistemology, the investigation of information and conviction, is affected by the algorithmic universe hypothesis. On the off chance that the universe works as per exact calculations, our insight and comprehension of it very well may be viewed as a work to translate these codes. This viewpoint challenges customary ideas of information and conviction, stressing the cutoff points and capability of human comprehension.

3. Philosophy of Mind: The way of thinking of brain investigates the idea of cognizance and mental states. The algorithmic universe hypothesis recommends that cognizance and mental cycles are administered by basic calculations. This view can impact banters about the idea of the psyche, the connection among brain and body, and the chance of counterfeit awareness.

4. Philosophy of Science: The way of thinking of science looks at the standards and techniques basic logical request. The algorithmic point of view lines up with the possibility that logical hypotheses and models are endeavors to grasp the central calculations of the universe. This view underlines the job of numerical and computational strategies in logical revelation.

Moral Ramifications of the Algorithmic Universe

The idea of the universe as an algorithmic framework has significant moral ramifications, impacting how we might interpret ethical quality, obligation, and human way of behaving:

1. Determinism and Free Will: Assuming the universe works as per deterministic calculations, inquiries regarding through and through freedom and moral obligation emerge. The algorithmic viewpoint challenges the idea of outright through and through freedom, proposing that our activities and choices might be affected by hidden examples and codes.

2. Ethical Standards and Algorithms: Moral standards can be seen as calculations that guide human way of behaving and direction. The algorithmic universe hypothesis proposes that virtues and moral standards might reflect further examples and codes administering human cooperations and cultural designs.

3. The Job of Intuition: Instinct assumes a critical part in moral navigation. The algorithmic point of view can reveal insight into how instinctive decisions line up with hidden

calculations and examples. Understanding these associations can improve our capacity to pursue moral choices.

4. Artificial Knowledge and Ethics: The improvement of man-made intelligence and AI acquaints moral contemplations related with algorithmic independent direction. The algorithmic universe hypothesis underscores the significance of adjusting man-made intelligence frameworks to moral standards and guaranteeing that they work in manners that reflect human qualities and profound quality.

Contextual investigations and True Applications

To outline the effect of the algorithmic universe hypothesis on way of thinking and morals, we should look at a couple of contextual investigations and true applications:

1. Predictive Policing: Prescient policing utilizes calculations to conjecture crime and apportion policing. This application raises moral worries about decency, inclination, and the expected effect on common freedoms. The algorithmic point of view features the requirement for straightforward and responsible algorithmic frameworks in policing.

2. Autonomous Vehicles: Independent vehicles depend on calculations to arrive at continuous conclusions about driving. Moral situations emerge in regards to how these frameworks ought to focus on wellbeing and handle situations including likely mischief. The algorithmic universe hypothesis highlights the significance of adjusting independent frameworks to moral qualities and cultural standards.

3. Ethical man-made intelligence Design: The plan of moral man-made intelligence includes making calculations that reflect human qualities and moral standards. By understanding the

algorithmic idea of the universe, analysts can foster rules and structures for planning simulated intelligence frameworks that advance reasonableness, straightforwardness, and responsibility.

4. Algorithmic Bias: Algorithmic predisposition alludes to the accidental separation or unreasonable treatment coming about because of one-sided calculations. Tending to algorithmic predisposition requires understanding and relieving the elements that impact algorithmic navigation, guaranteeing that these frameworks work in manners that line up with moral standards.

Suggestions for Theory and Morals

The algorithmic universe hypothesis has a few ramifications for reasoning and morals:

1. Revisiting Philosophical Foundations: The algorithmic point of view empowers a reevaluation of customary philosophical establishments. By incorporating calculations into how we might interpret reality, we can foster new structures for resolving basic inquiries regarding presence, information, and ethical quality.

2. Ethical Systems for Technology: The turn of events and execution of innovation should be directed by moral structures that line up with the algorithmic standards of the universe. This approach guarantees that innovative progressions advance human prosperity and reflect cultural qualities.

3. Enhancing Moral Understanding: Understanding moral standards as calculations can improve our appreciation of moral way of behaving and navigation. This viewpoint can prompt the advancement of additional successful techniques for advancing moral lead and settling moral problems.

4. Philosophical Integration: Incorporating the algorithmic universe hypothesis with conventional philosophical idea can prompt a more extensive comprehension of the real world. This joining energizes interdisciplinary exchange and joint effort, advancing our philosophical and moral points of view.

The Eventual fate of Reasoning and Morals

As we keep on investigating the algorithmic universe hypothesis, the eventual fate of reasoning and morals holds energizing prospects:

1. Algorithmic Morals Research: Continuous examination into algorithmic morals will propel how we might interpret how calculations can be planned and carried out in manners that reflect moral standards. This examination will add to the advancement of capable and impartial mechanical arrangements.

2. Philosophical Investigation into AI: The way of thinking of simulated intelligence will keep on developing, resolving inquiries regarding the idea of fake cognizance, the job of calculations in navigation, and the moral ramifications of simulated intelligence frameworks. This request will extend how we might interpret the connection among innovation and reasoning.

3. Ethical Systems for Arising Technologies: As new innovations arise, moral structures should be created to address their extraordinary difficulties and suggestions. The algorithmic viewpoint will assume a vital part in molding these systems and guaranteeing that mechanical headways line up with moral qualities.

4. Global Moral Standards: The improvement of worldwide moral guidelines for innovation and calculations will be fundamental for tending to multifaceted and global difficulties. By coordinating the algorithmic universe hypothesis, we can lay out rules that advance

decency, straightforwardness, and regard for assorted points of view.

CONCLUSION:

Embracing the Algorithmic Universe

Billy Carson's investigation of the algorithmic universe offers a historic system that spans science, innovation, reasoning, and human expression. By recommending that the universe works as indicated by central calculations, Carson gives a bringing together viewpoint that extends how we might interpret reality and our place inside it. This end ponders the extraordinary capability of the algorithmic universe hypothesis and its suggestions for different spaces of human request and movement.

The Binding together Force of Calculations

Carson's hypothesis features the significant interconnectedness of all peculiarities through hidden calculations. This point of view rises above conventional disciplinary limits, uncovering that the universe works as a cognizant framework represented by numerical and computational standards. By perceiving the universe's algorithmic nature, we can accomplish a more all encompassing comprehension of different fields, from cosmology to cognizance, from craftsmanship to morals.

Propels in Science and Innovation

The algorithmic universe hypothesis has huge ramifications for logical and innovative progressions. By review the universe as a complicated calculation, we can foster more exact models and reproductions of regular cycles. This approach upgrades our capacity to investigate the universe, figure out natural frameworks, and develop in fields, for example, computerized reasoning and quantum processing. The utilization of algorithmic standards vows to drive future revelations and innovative forward leaps, cultivating a more profound arrangement among science and innovation.

Improving Wellbeing and Prosperity

In the domain of wellbeing and prosperity, the algorithmic universe hypothesis offers new bits of knowledge into the working of the human body and brain. By understanding wellbeing as administered by complex calculations, we can foster more viable clinical medicines, emotional well-being mediations, and comprehensive wellbeing procedures. This point of view energizes customized medication, integrative wellbeing draws near, and inventive treatments that line up with the crucial standards of equilibrium and amicability.

Changing Craftsmanship and Inventiveness

Carson's hypothesis additionally upsets our way to deal with craftsmanship and imagination. By perceiving the algorithmic idea of creative cycles, we can investigate new types of articulation and advancement. The reconciliation of calculations and innovation in workmanship improves imaginative conceivable outcomes and opens up clever roads for creative investigation. This viewpoint improves our enthusiasm for craftsmanship and highlights the profound association among imagination and computational standards.

Rethinking Theory and Morals

The algorithmic universe hypothesis challenges and advances conventional philosophical and moral structures. By survey reality from the perspective of calculations, we gain new experiences into key inquiries regarding presence, information, and ethical quality. This viewpoint illuminates moral direction, especially with regards to innovation and man-made brainpower, and supports the advancement of capable and fair structures for exploring complex moral predicaments.

The Way ahead

As we push ahead, embracing the algorithmic universe hypothesis offers a way to more profound comprehension and development. The combination of this point of view into different fields can prompt groundbreaking headways, from logical exploration to imaginative creation, from medical care to moral practice. By recognizing the algorithmic idea of the universe, we open ourselves to additional opportunities and approaches that line up with the essential codes overseeing presence.

Last Contemplations

Billy Carson's algorithmic universe hypothesis addresses a change in perspective that welcomes us to investigate the universe from the perspective of mind boggling calculations and numerical examples. This point of view upgrades how we might interpret the universe as well as gives commonsense bits of knowledge and advancements across different areas. As we proceed to investigate and apply these standards, we move towards a future where science, innovation, reasoning, and inventiveness unite, prompting a more extravagant and more interconnected comprehension of the world.

All in all, embracing the algorithmic idea of the universe offers a significant chance to propel human information and work on our lives. By coordinating Carson's experiences into our requests and practices, we can open new components of understanding and make an additional agreeable and smart world.

APPENDIX:

Supplementary Material and Resources

The index gives extra assets, references, and beneficial material to help the experiences and ideas introduced all through the book. It incorporates a determination of key readings, devices, and philosophies connected with the algorithmic universe hypothesis and its applications.

A. Glossary of Terms

Algorithm: A bunch of rules or directions intended to play out a particular errand or tackle an issue, frequently executed in PC programming or numerical computations.

Epistemology: The part of reasoning worried about the nature and extent of information, including its cutoff points and legitimacy.

Fractals: Mathematical examples that are self-comparable across various scales and can be depicted by iterative calculations. They are in many cases used to demonstrate perplexing, normal peculiarities.

Generative Art: Craftsmanship made utilizing calculations or computational cycles, where the last fine art rises up out of the guidelines and boundaries set by the craftsman.

Neuroplasticity: The mind's capacity to rearrange itself by framing new brain associations, frequently because of learning or injury.

Prescient Policing: The utilization of calculations and information examination to estimate crime and allot policing in like manner.

Quantum Computing: A kind of registering that uses the standards of quantum mechanics to handle data in manners that conventional PCs can't, possibly tackling complex issues all the more productively.

Manufactured Biology: The plan and development of new organic parts, gadgets, and frameworks, or the overhaul of existing, normal natural frameworks.

B. Key Readings and References

1. Carson, B. (2023). The Algorithmic Universe: Unraveling the Infinite Code. Distributer.

2. Gleick, J. (2011). The Data: A Set of experiences, a Hypothesis, a Flood. Pantheon Books.
 - An investigation of how data hypothesis and calculations shape how we might interpret the world.

3. Mandelbrot, B. B. (1982). The Fractal Math of Nature. W.H. Freeman and Company.
- An original work on fractals and their application in figuring out regular examples.

4. Pinker, S. (2002). The Clean canvas: The Advanced Forswearing of Human Nature. Viking Penguin.
- A conversation of human instinct, insight, and the ramifications for grasping way of behaving and inventiveness.

5. Reas, C., and Sear, B. (2007). Processing: A Programming Handbook for Visual Fashioners and Artists. MIT Press.
- An extensive manual for generative craftsmanship utilizing the Handling programming language.

6. Vinge, V. (1993). The Coming Mechanical Peculiarity: How to Make due in the Post-Human Era. Procedures of the Vision-21: Interdisciplinary Science and Designing Gathering.

- A critical work on the ramifications of cutting edge innovation and computerized reasoning.

C. *Instruments and Procedures*

1. Algorithmic Plan Tools
- Processing: An adaptable programming sketchbook and a language for figuring out how to code inside the setting of the visual expressions.
- p5.js: A JavaScript library that makes coding illustrations and intuitive substance simple for specialists and planners.

2. Information Investigation and Visualization
- R Programming: An open-source language and climate for factual processing and illustrations.
- Tableau: A useful asset for making intelligent information perceptions and dashboards.

3. Wellbeing and Health Apps

- MyFitnessPal: An application for following sustenance and exercise, integrating calculations to customize wellbeing proposals.

- Headspace: A care application that utilizations directed reflection methods to work on mental prosperity.

4. Moral simulated intelligence Frameworks
- Simulated intelligence Morals Rules by the European Commission: A structure for creating and executing computer based intelligence frameworks that line up with moral norms.

- The Asilomar simulated intelligence Principles: Rules for the improvement of man-made consciousness to guarantee its positive effect on society.

D. Extra Assets

1. Online Courses
- Coursera: Offers seminars on information science, algorithmic plan, and man-made reasoning from top colleges.

- edX: Gives admittance to seminars on way of thinking, morals, and the effect of innovation on society.

2. Research Journals
- Diary of Man-made consciousness Research: Distributes research on artificial intelligence calculations and their applications.
- Reasoning and Technology: A diary investigating the philosophical and moral ramifications of innovation.

3. Proficient Organizations
- IEEE PC Society: A main association for software engineering and designing experts, zeroing in on calculations and innovation.
- American Philosophical Association: An association devoted to advancing philosophical examination and exchange.

E. Affirmations

This addendum is planned to give perusers the devices and assets important to additionally investigate the ideas examined in the book. It is critical to recognize the commitments of analysts, designers, and scholars whose work has impacted the comprehension of calculations, the universe, and their convergences with human request.

ACKNOWLEDGEMENTS

The finishing of this book, The Algorithmic Universe: Billy Carson's Investigation of Vast Codes, addresses a huge undertaking, and I'm thankful to the people who have upheld and added to this task.

Billy Carson: My most profound appreciation goes to Billy Carson for his spearheading experiences into the algorithmic idea of the universe. His work has been the foundation of this book, offering a progressive point of view that has motivated and directed the investigation of this subject.

Research and Scholars: I stretch out my appreciation to the scientists and researchers whose works have given significant setting and profundity to the conversation of calculations, cosmology, and related fields. Their commitments have been instrumental in forming the thoughts introduced in this book.

Scholastic and Expert Advisors: To the scholar and expert counselors who offered their mastery, input, and consolation all through the creative cycle, your direction has been important. Your help has refined the ideas and guarantee the exactness and pertinence of the material.

Editors and Reviewers: Unique because of the editors and commentators for their careful scrupulousness and useful input. Your endeavors have essentially upgraded the clearness and nature of the book.

Family and Friends: I'm significantly thankful to my loved ones for their faithful help and understanding. Your support has been a wellspring of solidarity and inspiration all through this excursion.

Readers: At last, I need to thank the perusers for their advantage in investigating the algorithmic universe. Your commitment with the thoughts introduced in this book at last makes this work significant.

This book is a cooperative accomplishment, based upon the commitments and backing of numerous people. It is my expectation that the experiences and points of view shared here will move further investigation and comprehension of the algorithmic idea of the universe.

www.ingramcontent.com/pod-product-compliance
Lightning Source LLC
Chambersburg PA
CBHW071939210526
45479CB00002B/741